另一面的阳光

Sun on Both Sides

Gunter Pauli

冈特·鲍利 著

唐继荣 译

www.xuelinpress.com

丛书编委会

主　任：贾　峰

副主任：何家振　郑立明

委　员：牛玲娟　李原原　李曙东　吴建民　彭　勇
　　　　冯　缨　靳增江

丛书出版委员会

主　任：段学俭

副主任：匡志强　张　蓉

成　员：叶　刚　李晓梅　魏　来　徐雅清　田振军
　　　　蔡雩奇

特别感谢以下热心人士对译稿润色工作的支持：

姜竹青　韩　笑　杨　爽　周依奇　于　哲　阳平坚
李雪红　汪　楠　单　威　查振旺　李海红　姚爱静
朱　国　彭　江　于洪英　隋淑光　严　岷

不用多久,太阳不仅会照到叶子顶面,还会照到叶子底面

Soon the sun will be shining below as well as on top!

正在阳光下享受美好的一天

Enjoying a day in the sun

一株老鹳草正在南非桌山的阳光下享受美好的一天。一条毛毛虫从老鹳草的叶片上爬过,在其中一片叶子下方安顿下来,想打个盹。叶片下方是纳凉的好地方,能避开火辣辣的阳光,周围的鲜花也让这里成为一片美丽的休憩之处。

A geranium is enjoying a day in the sun on Table Mountain when a caterpillar crawls over one of its leaves and settles on the underside for a nap. The cool spot in the shade provides cover against the hot sun and the bright flowers make it a beautiful place for a rest.

目录

另一面的阳光	4
你知道吗?	22
想一想	26
自己动手!	27
学科知识	28
情感智慧	29
艺术	29
思维拓展	30
动手能力	30
故事灵感来自	31

Contents

Sun on Both Sides	4
Did you know?	22
Think about it	26
Do it yourself!	27
Academic Knowledge	28
Emotional Intelligence	29
The Arts	29
Systems: Making the Connections	30
Capacity to Implement	30
This fable is inspired by	31

"要是你认为待在那里又安全又凉爽,那你可得小心了!不用多久,太阳不仅会照到叶子顶面,还会照到叶子底面。"老鹳草提醒道。

"那不可能!"毛毛虫回答说。"只要太阳照在叶片上方,那么叶片底面总会有荫凉地儿。这是显而易见的。"

"If you think you can stay safe and cool there, beware, because soon the sun will be shining below as well as on top!" remarks the geranium.

"That cannot ever happen!" replies the caterpillar. "If the sun shines on top, then there will always be shade below. That is quite obvious."

"看上去似乎的确如此，但我身体下方的地上有一滩水，水会将阳光向上反射。一旦被反射的光线照到我的叶片底面，你就得享受一些可爱的温暖了。这就像人们使用的新型太阳能电池板一样。"

"享受？我失去了安全又凉爽的休憩处，怎么还能享受啊？"

"That may seem true, but there is a pool of water on the ground below me, and the water will reflect the sunlight back up. If the reflected light hits the bottom of my leaf you could enjoy some lovely warmth. It works just like with these new solar panels people use."

"Enjoy? How could I enjoy losing my safe and cool resting spot?"

水会将阳光向上反射

Water will reflect the sunlight back up

太阳能板只有顶面接收到阳光

Solar panels capture solar power only on top

"当然,你可能会不太高兴,是吧?或许你得再找别处休息了。但你相信吗,人们安放在屋顶用来捕获太阳能的所有太阳能板,只有顶面能照到阳光,底面从来接收不到阳光照射?"

"Well, you would not be too happy, would you now? You may have to find another resting spot. But can you believe that all these solar panels that people has been putting on their roofs to capture solar power only receive sun on top, never below?"

"是因为人们希望模拟自然吗?你们植物也只是利用其中一面的光能,对吧?"

"这与是否向大自然学习无关,更多地在于如何利用现有的资源。如果两个表面都能利用好,就能获得双倍的能量。"

"Is that because they want to imitate nature? You plants only use the sun on one side, don't you?"

"This has nothing to do with learning from nature. This has more to do with using what you have. By using both surfaces, you can get double the power."

如果两个表面都能利用好，就能获得双倍的能量

By using both surfaces, you can get double the power

我们应该尽可能多地利用现有的资源

We have to do more with what we have

"你是建议让这些用沙子做的太阳能板的两面都得到阳光的照射吗?"

"没错!如果你能从太阳能板的顶面和底面都获得能源,为什么只利用顶面呢?我们应该尽可能多地利用现有的资源。"

"So are you suggesting that these solar panels made from sand should have the sun shining on both sides?"

"Absolutely! If you could make energy on top and below, why would you only use the top? We have to do more with what we have."

"瞧，我喜欢荫凉地儿。没有荫凉地儿，就太热啦，甚至感觉要燃烧起来！而且我听说，如果太阳能电池变得过热，它们就不能正常工作了。"

"如果你热过头了，会怎样做呢？"

"Look, I like the shade. If there is no shade then it gets too hot. You could even burn up! And I have heard that if solar cells get too hot, they don't work so well."

"What do you do when it gets too hot?"

如果你热过头了，会怎样做呢？

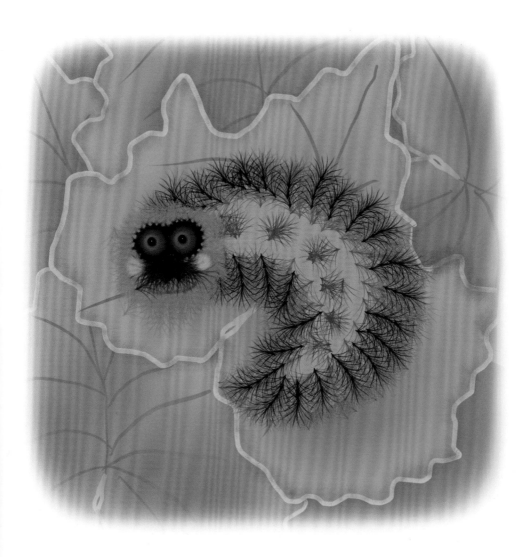

What do you do when it gets too hot?

凉爽又清新的身体

A cool and fresh body

"我？嗯，我喜欢去冲个澡。"

"对呀！如果太热，你会用水来降温。然后，你会得到什么？"

"凉爽又清新的身体，不是吗？"

"Me? Well, I like to take a shower."

"Exactly! When it is too hot, you use water to cool down, and then what do you have?"

"A cool and fresh body?"

"就这些吗?"老鹳草追问道。

"哦,我懂了!这样的话,房子就能获得更多能源,同时还有热水。这热水还可能很干净!这将是未来的生活方式。"

……这仅仅是开始!……

"Is that all?" wonders the geranium.

"Oh, now I get it. A house can have more energy and have hot water. It can even be clean, hot water! That is the way to live in the future."

... AND IT HAS ONLY JUST BEGUN!...

……这仅仅是开始!……

...AND IT HAS ONLY JUST BEGUN!...

Did You Know?

你知道吗?

Traditional solar panels only use the power of the sun on one side, and never on both sides.

传统的太阳能板只有一面能利用太阳能，从未两面同时利用。

Optics will allow the sun to shine directly on to the top of the panel, and shine in a concentrated way on the bottom.

光学设备将使得太阳不仅能直接照射太阳能板的顶面，而且能以一种聚合的方式照在它的底面。

If solar cells are kept at 50° C while exposed to the sun, they generate the most electricity.

暴露于阳光下的太阳能电池，当它们的温度保持在 50℃ 时，发电效率最高。

A car engine gets hot through combustion, but thanks to water-cooling, the engine keeps on working.

汽车发动机由于燃烧而变热，但水冷设备让它得以保持运转。

𝒯iny water pipes inside a photovoltaic panel maintain the panel at a constant temperature, while generating hot water.

光伏电池板内的微小水管让电池板保持温度恒定，同时产生热水。

𝒮olar energy streams like a direct current (DC), while our electric network operates on the alternate current (AC) standard.

太阳能像直流电一样流动，而我们的电力网络系统则是按交流电的标准运行的。

Chargers for computers, toys and games do not only charge, they also convert the electricity from AC (the power of the grid) to DC (the power system of solar panels).

计算机、玩具和游戏机的充电器不仅仅能用于充电，还能将电流从电网系统中的交流电转换为太阳能板系统里的直流电。

A solar panel made of sand generates more energy (thermal) from heating and cooling water than from the electricity (power) it generates.

用沙子做成的太阳能板，通过加热和冷却水产生的热能比它发电产生的能量还要多。

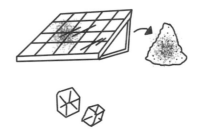

Think About It 想一想

How competitive would solar panels be if they generated double the amount of energy, some as electricity and some as heat?

如果太阳能板产生的能量可以翻倍，其中电能和热能各占一部分，它的竞争力会如何？

工程师们会怎么看这个问题，是将光伏电池变得更薄更轻来节约能源，还是将太阳能电池变厚来获得双倍的能量？

How might engineers look at making photovoltaic cells thinner and lighter to save energy, while thicker cells generate double the revenue?

随着专业化程度的加深，太阳能专家中有的专注于发电，有的专注于产生热水，这是否会导致人们最终购买两套太阳能电池板系统，而不是一套？

Does specialisation, with one set of solar experts for electricity and another set of solar experts for hot water, have the effect that people end up buying two systems instead of one?

处于高热状态时，怎样降温才是最节约能源的？

What are energy efficient ways to cool down when it is hot?

Do It Yourself !
自己动手！

Take a piece of black metal (if you can't find black metal, then paint it black). It should be at least 10 cm long and 2 cm wide, but the bigger the better. Then find some aluminium foil that will reflect the sun from an open space above, onto the bottom of the black metal plate. Find out, through trial and error, firstly the right angle and secondly the best curve of the aluminium foil so that the sunlight is reflected onto the bottom surface of the panel for the longest possible time. You should design the foil so that the sun is concentrated on the bottom surface for at least four hours. It may be a good idea to work in four or five teams and share your experiences to find the best solution faster. It may not be easy, but it will be fun!

找一块黑色金属（如果没有，就把其他金属涂成黑色）。尺寸应不小于10厘米长、2厘米宽，越大越好。找些能向上反射阳光的铝箔，使它们能在开阔地方将阳光反射到黑色金属板的底面。通过反复试验，首先找到正确的角度，其次是铝箔的最佳弧度，从而令阳光反射到黑色金属板底面的时间最长。设计好铝箔后，确保阳光能集中照射到黑金属板底部至少有4小时。不妨分为4~5组分别进行试验，大家分享经验，从而更快找到最佳解决方案。这可能不太容易，但会很有趣。

TEACHER AND PARENT GUIDE

学科知识
Academic Knowledge

生物学	南非丰博斯植被的生物圈；历史上曾将老鹳草与天竺葵类植物的名称混淆。
化 学	如何将沙子（硅）转化为石英（二氧化硅）；用盐酸和三氯硅烷来提纯石英；与铅有关的健康危险。
物 理	光伏电池发电的理想温度；通过紫外线和加热来净化水；水中的热传递；用热泵从冷源产热，或从热源制冷。
工程学	如何把同一个装备中的热水生产与能源生产相结合；将基于某类工业标准（汽车）上的技术和创新集群应用到另一类工业（太阳能）上；如何将剧毒和能源密集的太阳能电池生产过程转变为可持续的过程。
经济学	生产力计划；用你现有的资源生产更多的东西；关注核心竞争力，要把那些显而易见但在中心范围之外的机遇创造性地转化成核心业务。
伦理学	做得不那么糟糕，仍然是糟糕；若有机会做得更好却拒绝去尝试，那是真的很糟糕；如果一个微小的改变就让它们变得非常具有竞争力，那么还能认为有必要对它进行补贴吗？
历 史	镜子最早于公元前8000年在土耳其使用；希腊人知道凹凸镜的特性；亚历山大·贝克勒尔是首位观察到光伏效应的人（1839年）；1894年，首个太阳能电池专利被授予梅尔文·西弗里。
地 理	南非桌山和好望角；地球哪个部分接收到最多的阳光照射？北极的太阳能产业会有竞争力吗？
数 学	几何光学；反射（物理学）与折射（数学）。
生活方式	我们接受了只利用太阳能板一面的标准做法，从不质疑它的效率；对于单个的光伏电池，我们除了希望获得更多的电能输出之外，也没有强制要求它表现出其他方面的性能。
社会学	社会对可再生能源的接受较慢，似乎需要政府干预。
心理学	心理性近视：我们被培养成只关注我们自己希望和需要的事物，对其他任何不符合我们固有逻辑的观点却缺乏思考的能力和胸怀。
系统论	我们还处于脱离实际的学术象牙塔中，只进行着非常集中但相互隔绝的科学研究。

教师与家长指南

情感智慧
Emotional Intelligence

毛毛虫

毛毛虫知道自己希望得到什么——只有荫凉而已，他不相信任何有可能威胁到自己的安乐窝的事物。他只依据自身的利益来考虑问题，对其他现实情况视而不见。老鹳草迫使毛毛虫去观察自己的降温行为。这有助于激励毛毛虫去聆听、思考，并超越过去的思维定势。毛毛虫通过亲身体会，理解了老鹳草一直在设法分享的逻辑和思想。由于获得了新的见解，他现在不仅被说服了，甚至变得很热心。

老鹳草

老鹳草处于一种静坐不动的生活状态。它知道自己无法移动，因此花时间观察周围的各种现实情况。它对毛毛虫抱有同情心，警告毛毛虫不久将失去阴凉。当毛毛虫表示难以置信时，老鹳草花时间解释了为什么毛毛虫所处的位置不能长时间保持阴凉。老鹳草扩展了这一讨论主题，进行了更深入的独到反思。它将讨论引向了更广泛的话题，这种思考方式很有借鉴意义，甚至还提出一些称得上人生观的核心原则。老鹳草随后从解释转为提问，质疑毛毛虫的行为。这种方式不仅吸引了毛毛虫的全部注意力，而且帮助老鹳草引导毛毛虫获取新信息，最终让毛毛虫形成了对事物新的积极态度。

艺术
The Arts

找到一个你想用作手绘目标的物体，先以它为模特绘制一幅简单的素描。然后，通过凸透镜观察这个物体，并将在镜子中看到的景象画下来。再通过凹透镜观察同一个物体，并将镜子中看到的景象画下来。最后将三幅画放在一起，你会观察到什么现象？

TEACHER AND PARENT GUIDE

思维拓展
Systems: Making the Connections

　　太阳每天都在发光。即便云层很厚，仍然可以看到太阳光。据说只需到达地球土壤和水体的太阳总能量的1%，就可以满足我们的全部能源需求。然而，不可能把整个地球都盖上太阳能设备，所以我们需要用最有效率的方法来处理此事。当我们应用光学技术将太阳光汇聚到太阳能板底部时，若能把微小的水管置于太阳能板内部来给光伏电池降温，那么为了容纳这些小管，这种太阳能板会比正常的太阳能板更厚。从发电的角度来说，这将增加原材料的成本。但另一方面，这种太阳能电池"三明治"中的小管，却能够以非常低的成本产生热水。这种综合考虑热水和电力的技术，将从太阳光中捕获更多的能量，而额外的材料成本却远低于新增的能量价值。因为高热可以杀死水中的细菌，所以水一旦被加热，同时也会被净化。为了在太阳能电池底部产生聚焦能量的镜面效果，在太阳能板和反射装置的底部之间有20厘米的空隙，这个空间充满了空气。如果将太阳能系统用作建筑的实际屋顶，而不是放置在屋顶上，这个空间就可以当作隔热层，帮人们省去购买保温或御寒的特殊绝缘材料的花费。实际上，最初的太阳能电池已经逐渐发展成一个提供许多好处的系统，包括电能、热水、用作屋顶、隔热和净化水质。有了可再生能源，房主能节约资金去进行其他投资，也能确保我们的生活品质更高更健康。

动手能力
Capacity to Implement

　　你们学校使用太阳能吗？你们学校使用任何形式的可再生能源吗？为什么不请求约见学校校长，询问给建筑加热和降温需要多少能源，以及全年的照明需要多少电力？或许校长会告诉你用于加热、降温和照明的总预算。如果学校所有建筑的屋顶都装上光伏电池做成的太阳能板，估算这将产生多少电力。然后，如果所有可利用的屋顶空间都装上太阳能战士公司的太阳能板，将产生多少电力和热量。学校是否有足够的屋顶空间来进行太阳能利用，使得学校可以不依赖外部电网？请列出你的论据，系统阐述你的看法，然后看能否说服校长。

故事灵感来自

斯特凡·拉松
Stefan LARSSON

斯特凡·拉松学习的专业是计算机科学和电气工程。他曾经在太阳能和生物燃料领域从事过研究工作，并成为瑞典达拉纳技术高中工业管理学部的助理教授。他职业生涯的很大一部分时间用于在瑞典和德国的工业能源巨头大瀑布电力公司（Vattenfall）研究聚光太阳能。当大瀑布电力公司决定不把太阳能作为其未来战略的一部分后，他离开了该公司。然后，他与尼古拉斯·斯腾隆德一起创办了太阳能战士公司(Solarus)，目标是研发、生产和销售聚光太阳能技术，通过将电能和热能结合在一起，使太阳能更具竞争力。他将第一个试点项目放在了自己的祖国瑞典。他意识到，如果太阳能在靠近北极圈的地方都具有竞争力，那么它将在地球任何地方都具有竞争力。拉松教授是瑞典聚光太阳能研究基金会的董事会成员。

更多资讯

www.solarus.se

http://www.pv-magazine.com/archive/articles/beitrag/the-trouble-with-silicon-_100001055/86/?tx_ttnews%5BbackCat%5D=115&cHash=5581693680d6204f040e6c4471f69c6b#axzz34F7G4FNX

图书在版编目（CIP）数据

另一面的阳光：汉英对照／（比）鲍利著；唐继荣译．－－上海：学林出版社，2015.6
（冈特生态童书．第2辑）
ISBN 978-7-5486-0874-5

Ⅰ．①另… Ⅱ．①鲍… ②唐… Ⅲ．①生态环境－环境保护－儿童读物－汉、英 Ⅳ．① X171.1-49

中国版本图书馆 CIP 数据核字（2015）第 092485 号

————————————————————————————

© 2015 Gunter Pauli
著作权合同登记号 图字 09-2015-446 号

冈特生态童书
另一面的阳光

作　　者——	冈特·鲍利
译　　者——	唐继荣
策　　划——	匡志强
责任编辑——	李晓梅
装帧设计——	魏　来
出　　版——	上海世纪出版股份有限公司 学林出版社
	地　址：上海钦州南路81号　电话／传真：021-64515005
	网址：www.xuelinpress.com
发　　行——	上海世纪出版股份有限公司发行中心
	（上海福建中路193号 网址：www.ewen.co）
印　　刷——	上海图宇印刷有限公司
开　　本——	710×1020　1/16
印　　张——	2
字　　数——	5万
版　　次——	2015年6月第1版
	2015年6月第1次印刷
书　　号——	ISBN 978-7-5486-0874-5/G·323
定　　价——	10.00元

（如发生印刷、装订质量问题，读者可向工厂调换）